Tables in Context

Patricia Westmoreland

Edward Arnold

Introduction

Learning tables can be a chore, but trying to cope without them in schoolwork or in real life situations can be difficult.

The more you practise, the easier they will become.

This book is set out in sections, starting with the two times table. Make sure you know your table before you begin each section.

I hope the puzzles, games and problems will make you want to know your tables – miracles sometimes do happen!

P.A.W.

© Patricia Westmoreland 1985

First published in Great Britain 1985
by Edward Arnold (Publishers) Ltd
41 Bedford Square
London WC2 3DQ

Edward Arnold (Australia) Pty Ltd
80 Waverley Road
Caulfield East 3145
PO Box 234
Melbourne

British Library Cataloguing in Publication Data

Westmoreland, Patricia
   Tables in context.
   1. Multiplication—Tables—Juvenile literature
   I. Title
513'.2'0212   QA49
ISBN 0-7131-8252-0

An answers leaflet is available on application to the publisher.

All rights reserved. No part of this publication may be reproduced, stored in a retrieval system, or transmitted in any form or by any means, electronic, photocopying, recording, or otherwise, without the prior permission of Edward Arnold (Publishers) Ltd

Acknowledgement
The publishers would like to thank Flamingo Land Zoo and Holiday Village for permission to use the brochure on p. 18.

Text set in 12/13 Souvenir
by Oxprint Ltd, Oxford
Printed in Great Britain at The Pitman Press, Bath

Contents

1 Who's our Mum? (× 2, ÷ 2) — 4
2 Limpon's Shoe Shop (× 2) — 5
3 Going camping (× 3) — 6
4 The sweet shop (÷ 3, × 3) — 7
5 Inter-House Competition (Revision: ÷ 2, ÷ 3) — 8
6 What! Four of them? (× 4) — 9
7 BMX bike race (÷ 4, × 4) — 10
8 Space Ball (× 5) — 11
9 The School Fair (÷ 5) — 12
10 Bob-a-Job Week (Revision: × 5, ÷ 5) — 13
11 Hit for six! (× 6) — 14
12 Rumblings at sea (÷ 6) — 15
13 The Garden Centre (Revision) — 16
14 Brian Bevan – Disc Jockey (Revision) — 17
15 The Zoo (× 7) — 18
16 Battery hens (÷ 7, ÷ 6) — 20
17 Wordwalls (Revision) — 21
18 Run, postman, run (× 8) — 22
19 The Mad Inventor (÷ 8) — 23
20 Sponsored event (× 9) — 24
21 The School Play (÷ 9) — 25
22 10p Sale (× 10) — 26
23 The Town Band (÷ 10) — 27
24 Three pints today, please! (Revision) — 28
25 Assault course (Revision: × 7, × 8, × 9) — 29
26 Wordwalls (Revision) — 30
27 Coded messages (Revision) — 31

1 Who's our Mum?

Chris & Rebecca

These proud mums have all had twins.

1. To find the mother of each pair of twins, multiply the number on their balloon by two.
   Write down your working out with the names of the mothers and children.

How old are the twins?

2. The ages of Dave and Patti add up to twenty years. How old are they? (Divide the total by two.)

3. The ages of Melanie and Emma add up to eighteen years. How old are they?

4. The ages of Brian and Peter add up to fourteen years. How old are they?

5. The ages of Daniel and Damien add up to eight years. How old are they?

6. The ages of Chris and Rebecca add up to sixteen years. How old are they?

7. The ages of Kathryn and Jane add up to four years. How old are they?

2 Limpon's Shoe Shop

Miss Biggs has to put these shoes away before closing time.
If you multiply the size of the shoe by two you will get the style number.

1. Make a list beginning like this:
   $2 \times 2 = 4$ The size 2 shoes are style 4.

How much?

The prices of the shoes are shown on the shoe boxes.

2. Style 20 shoes cost £9 a pair. How much do two pairs cost?
3. How much do two pairs of style 6 shoes cost?
4. Style 10 shoes cost £2 a pair. How much do eight pairs cost?
5. How much do two pairs of style 16 shoes cost?
6. How much do two pairs of style 14 shoes cost?
7. How much do four pairs of style 12 shoes cost?
8. How much do two pairs of style 18 shoes cost?

3 Going camping

```
                                              22nd May.

Dear Parent,
        Your child will require the following items for
the school camping trip to Whitby on 17th June.

    1 sleeping bag       4 blankets          2 pairs of boots
    6 pairs of socks     3 jumpers           3 pairs of trousers
    9 tins of beans      8 tins of soup      7 bars of chocolate
    7 packets of Smash   5 tins of rice pudding
                        Yours sincerely

                                F. Colley.
```

Tom, Dick and Henry are going on a trip. Copy and complete the table below to find out how much they need altogether of each item.

Equipment	Number	Total
sleeping bags	3 × 1	3
blankets	3 × 4	
pairs of boots	3 × 2	
pairs of socks		
jumpers		
pairs of trousers		
tins of soup		
tins of beans		
bars of chocolate		
packets of Smash		
tins of rice pudding		

4 The sweet shop

1. James spent 30p on choc bars. How many did he buy?
2. Alan spent 18p on mint lumps. How many did he buy?
3. Janet bought three fruity pops. How much did she spend?
4. Susan bought three bags of nuts. How much did she spend?
5. How much would three gobstoppers, three toffee sticks and three bags of nuts cost altogether?
6. How much would three mini drops, three choc bars and three bags of nuts cost altogether?
7. Which item would cost me 21p for three?
8. Which item would cost me 27p for three?
9. If I bought three of everything displayed in the window what would the total cost be?
10. You can spend up to a pound at the sweet shop, but you must buy things in threes. Write out a list of what you could buy and find the total cost.

5 Inter-House Competition

House Teams	POINTS TABLE		
	First	Second	Third
RED	36	10	7
BLUE	12	20	4
GREEN	15	18	8
YELLOW	18	6	8
	3 points for a first place	2 points for a second place	1 point for a third place

Red house scored 36 points for first places. Each first place scores 3 points. This means Red house gained 36 ÷ 3 = 12 first places.

1. How many first places did the other houses gain?
2. How many second places did each house gain?
3. Which house scored most points?
4. Which house scored fewest points?
5. Which house gained most places (first, second or third)?
6. Which house gained fewest places?
7. Last year Red house had five firsts and no thirds. Their total points were 25. How many second places did they have?
8. Last year Blue house had no thirds and eight seconds. Their total score was 37. How many firsts did they have?
9. There are several ways of scoring 18 points. How many can you list?

6 What! Four of them?

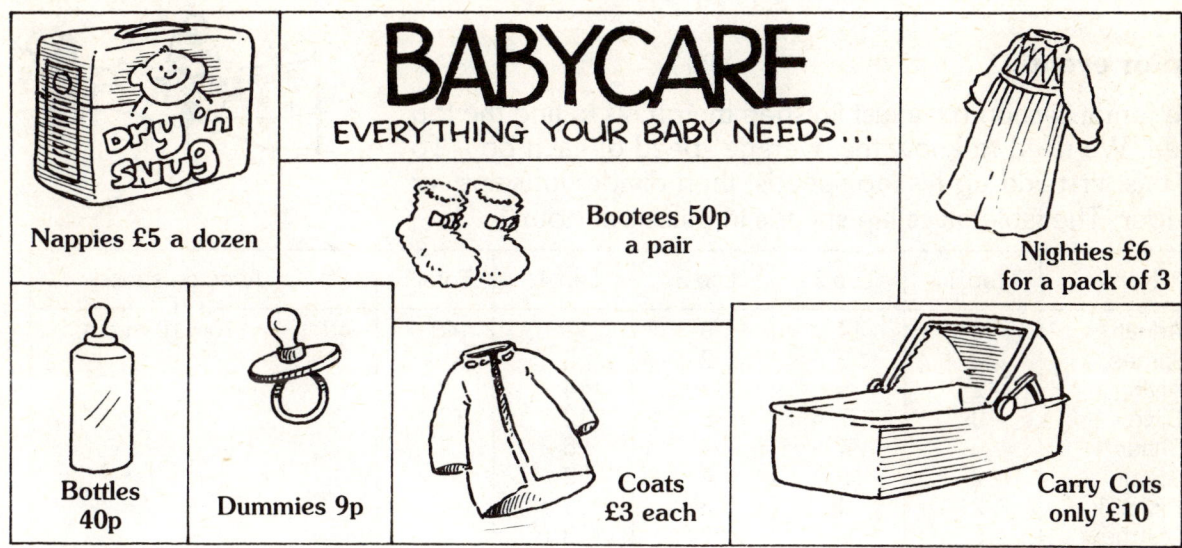

Mrs Baxter has just had quads so she needs to buy four sets of everything.

1. Copy and complete the table below to show how much she will have to pay.

Items needed for each baby	Cost for 1	Cost for 4
1 dozen nappies	£5	
1 pack of nighties		
1 carry cot		
1 dummy		
1 bottle		
1 pair of bootees		
1 coat		

2. The babies need feeding every four hours. Each baby has one bottle at each feed. How many bottles will one baby need in a day?

3. How many bottles will the four babies need in one day?

4. Work out the cost of buying the items in Question 1
   (a) for twins
   (b) for triplets.

7 BMX bike race

Junior event

The junior section have just finished their trials to find the top three. We need to know the average speed of each boy. To do this, first add up his lap speeds, then divide your answer by four. The table gives lap speeds in miles per hour.

Name	Lap 1	Lap 2	Lap 3	Lap 4	Total	Average speed
Adrian	6	14	9	11	40	40 ÷ 4 = 10 10 m.p.h.
James	3	5	2	6		
Alvin	12	5	7	4		
Clive	9	7	8	12		
Shaun	5	4	6	5		
Abdul	2	9	4	9		
Richard	2	4	3	3		
Matthew	9	8	7	8		

1. Write a list of the totals and average speeds for each boy.

2. List the three winners in order, starting with the one in first place.

3. What was the highest lap speed recorded?

4. What was the lowest lap speed recorded?

Senior event

5. The illustration represents the senior event. To find the speed that each boy is doing, multiply the number on his bike by four. Write a list starting like this:
Bike 5's speed is   5 × 4 = 20 m.p.h.

6. What is the average of the speeds of bikes 8, 7, 4 and 9? (Find the total speed, then divide by 4.)

7. Is the speed of bike 9 realistic?

8. Try to find out the highest speed recorded for a racing bike. Could a BMX bike rider reach that speed? Give reasons for your answer.

8 Space Ball

The Inter-Galactic match between planet Zendo and planet Unfoti is in progress.

1. Multiply the number on each player's ship by five to find the number of points he has scored. Make a list starting like this:
   Planet Unfoti
   $5 \times 7 = 35$ U7 scored 35 points

2. The goalie's number gives the number of goals he has saved. Multiply the number on each player's ship by five to find the

3. Add up the points scored by each team, including the points scored by the goalie. Write down the total score for each team.

4. Who won?

Match against Yenca

5. Last week the winning team made the same score against planet Yenca. Planet Yenca won the game by 5 points.
   If the goalie was Y3 and two of the players were Y7 and Y9, what could the other two players have been? (No two team members may have the same number.)
   Note: There is more than one solution to this question. How many can you find?

9 The School Fair

1. 50p was collected on 'Drown the Head'. How many people had a go?
2. 35p was made on the toffee apple stall. How many were bought?
3. 45p was made on the 'Wellie Throw'. How many people took part?
4. 25p was collected on 'Lift the Lady'. How many people tried?
5. 40p was collected by the fortune teller. How many people visited her?
6. 30p was collected at 'Test Your Strength'. How many people had a go?
7. Anne had 45p to spend. How many things could she try?
8. Nigel had 35p to spend. He wanted two goes on the 'Wellie Throw'. How many other things could he have a go on?
9. Andrew had 30p. Did he have enough to try everything?
10. If you had 60p to spend what would you try? Which would you go on more than once?

10 Bob-a-Job Week

	JOBS DONE
WASHING CARS	9
RUNNING ERRANDS	10
GARDENING	5
WASHING WINDOWS	4
DOG WALKING	8
FETCHING COAL	6

Blue Patrol did a summary of the jobs they had done.

1. How much did they earn for each type of job?
   Example: car washing   9 × 5p = 45p

2. How much did they earn altogether?

3. There are six members in Red Patrol. Altogether they earned £24. If they all earned the same amount, what would that be?

4. In Green Patrol, Harjit washed 8 cars, Stuart washed 6 and David washed 4.
   (a) How much did they earn each?
   (b) How much did they earn altogether?

5. In Red Patrol Alan ran 7 errands, Michael ran 3 and Carlton ran 5.
   (a) How much did they earn each?
   (b) How much did they earn altogether?

6. Yellow Patrol earned 30p for dog walking. How many jobs was this?

7. Green Patrol earned 60p for fetching coal. How many jobs was this?

8. Mrs Evans gave Harjit 50p. How many jobs should he do for her?

9. If Winston does two jobs a night, will he earn £1 in Bob-a-Job week?

10. Shaun earned £3.50. He did the same number of jobs each night.
    (a) How many jobs did he do each night?
    (b) How much did he earn each night?

11 Hit for Six!

Name	Number of Sixes					
J. Smailes	1	0	3	2	4	0
S. Lumsden	3	1	1	0	2	1
C. Palmer	1	1	1	0	0	1
D. Hall	0	0	0	3	2	0
R. Khan	2	2	0	2	2	1
K. Penrose	0	0	1	3	1	1
G. Field	2	1	0	0	2	2
D. Henderson	0	0	0	0	1	1
L. Miles	1	0	0	2	0	0
	30th April	7th May	21st May	28th May	4th June	11th June

The table shows the number of sixes hit by members of the school cricket team who are competing for the Hit for Six trophy.

1. (a) How many sixes did J. Smailes hit altogether?
   (b) What did he score altogether by hitting sixes?
   (c) What was his score on 4th June?

2. (a) How many sixes did S. Lumsden hit altogether?
   (b) What was his total score?
   (c) What was his score on 30th April?

3. What was C. Palmer's total score?

4. List the score totals for the rest of the competitors.

5. What was the highest score on 28th May and who scored it?

6. (a) In which week were the greatest number of sixes hit?
   (b) What was the total score gained by hitting sixes that week?

7. (a) In which week were the least number of sixes hit?
   (b) What was the total scored by hitting sixes that week?

8. Who won the trophy?

9. List the first three with their total scores.

12 Rumblings at sea

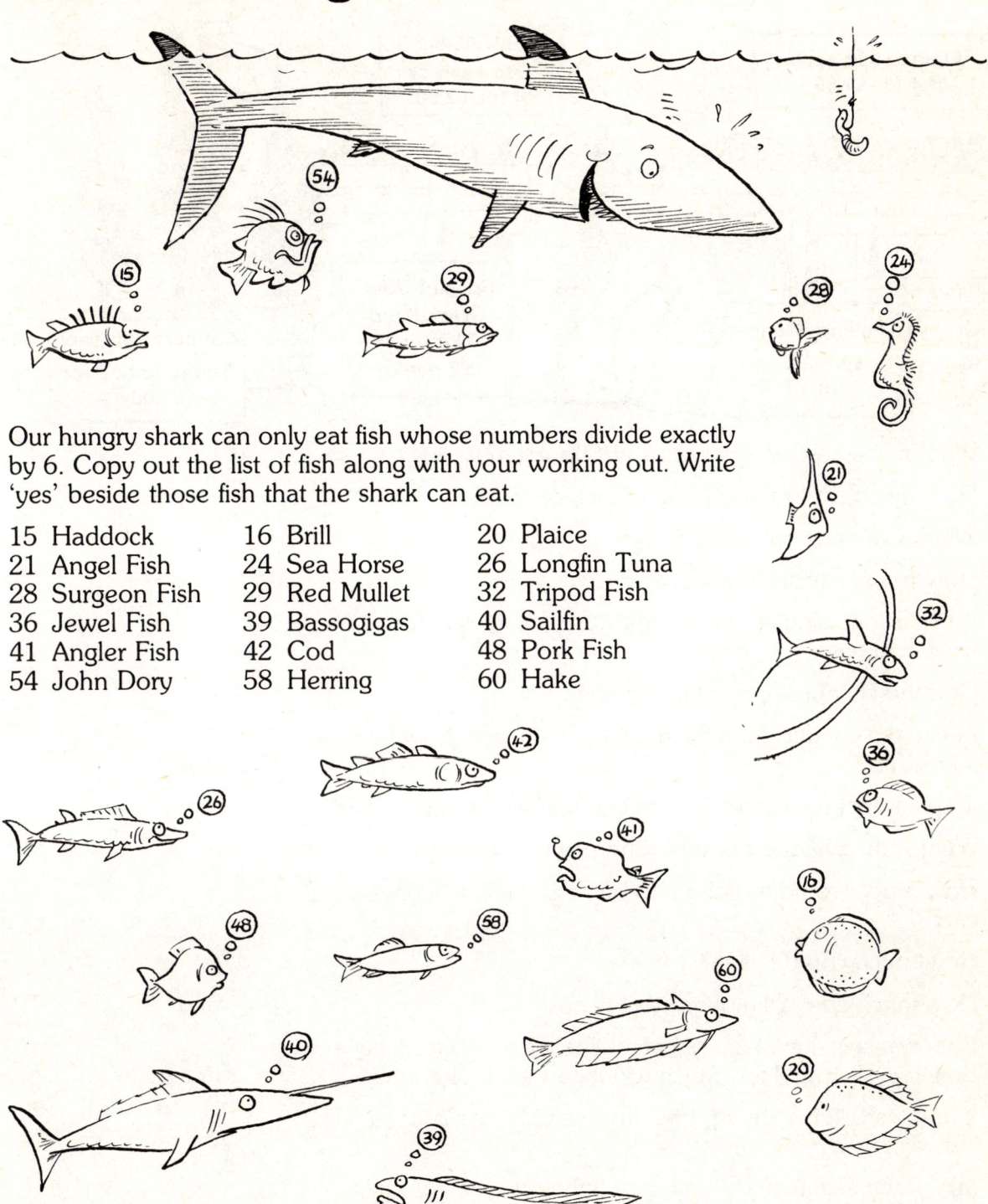

Our hungry shark can only eat fish whose numbers divide exactly by 6. Copy out the list of fish along with your working out. Write 'yes' beside those fish that the shark can eat.

15	Haddock	16	Brill	20	Plaice
21	Angel Fish	24	Sea Horse	26	Longfin Tuna
28	Surgeon Fish	29	Red Mullet	32	Tripod Fish
36	Jewel Fish	39	Bassogigas	40	Sailfin
41	Angler Fish	42	Cod	48	Pork Fish
54	John Dory	58	Herring	60	Hake

13 The Garden Centre

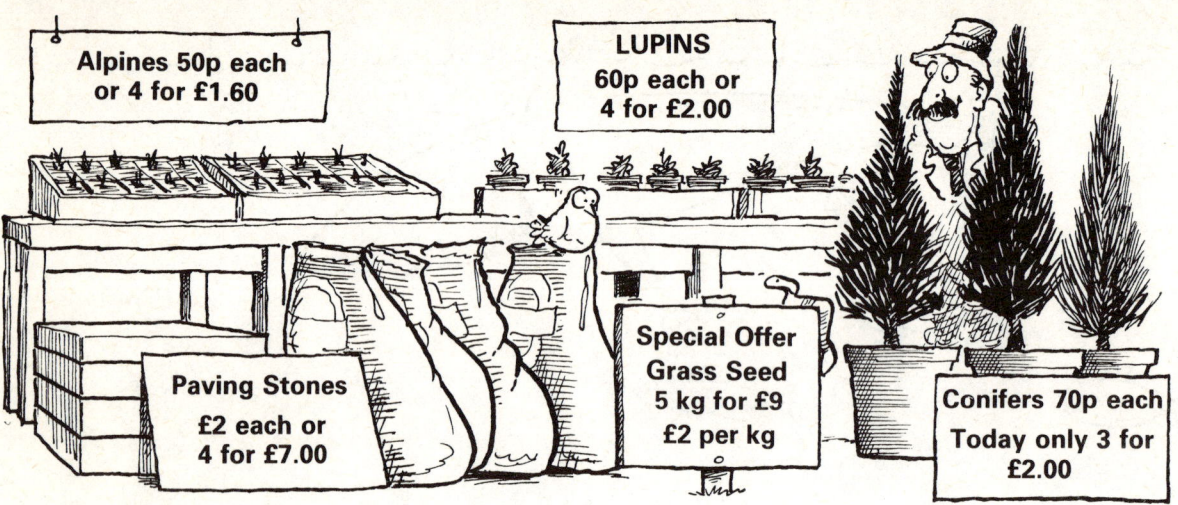

1. What is the cost of four lupins at 60p each?
2. How much do I save if I buy a pack of four?
3. What is the cheapest way to buy six lupins?
4. How much will six lupins cost me?
5. If I buy four alpines the cheaper way, how much will they have cost each?
6. Find the cheapest cost of ten alpines.
7. I need ten conifers for a hedge. If I buy them today what will it cost me?
8. If I wait until next week how much will ten conifers cost?
9. What is the cheapest cost of nine paving stones?
10. How much would it cost me to buy nine paving stones at £2 each?
11. How much could I save?
12. How much does 2 kg of grass seed cost?
13. Bob wants 2 kg of grass seed, Joe also wants 2 kg and Brian wants 1 kg. How much will it cost each of them?
14. If they shared a 5 kg bag how much could they save between them?
15. (a) How much would Brian save if they shared a 5 kg bag?
    (b) How much would Bob and Joe save each?

14 Brian Bevan – Disc Jockey

BOOKINGS DIARY

Day	Venue	Hours	
Monday	Rossiter's Night Club	4	£7 per hour
Tuesday	21st Birthday Party	6	£7 per hour
Wednesday	Hospital Broadcast	5	£4 per hour
Thursday	School Disco	3	£18 Total
Friday	Dinner Dance	6	Received £48
Saturday	Local Radio	2	Received £20
"	Rossiter's Night Club	4	£9 per hour

1. How much did Brian earn at Rossiter's on Monday?
2. How much did he earn there on Saturday?
3. How much did he earn in total at Rossiter's?
4. What did Brian earn at the 21st birthday party?
5. How much did he get for doing the hospital broadcast?
6. What was Brian's hourly rate for the disco? (Divide the fee he earned by the number of hours.)
7. What was Brian's hourly rate for
   (a) the dinner dance?
   (b) local radio?
8. What were his total earnings for Saturday?
9. Find Brian's total earnings for the week.
10. What was his highest hourly rate?
11. What was his lowest hourly rate?
12. Monday, Wednesday and Saturday are regular bookings. How much a week does Brian receive for these days altogether?
13. Why do you think that Brian's rates for Tuesday, Thursday and Friday vary so much?

15 The Zoo

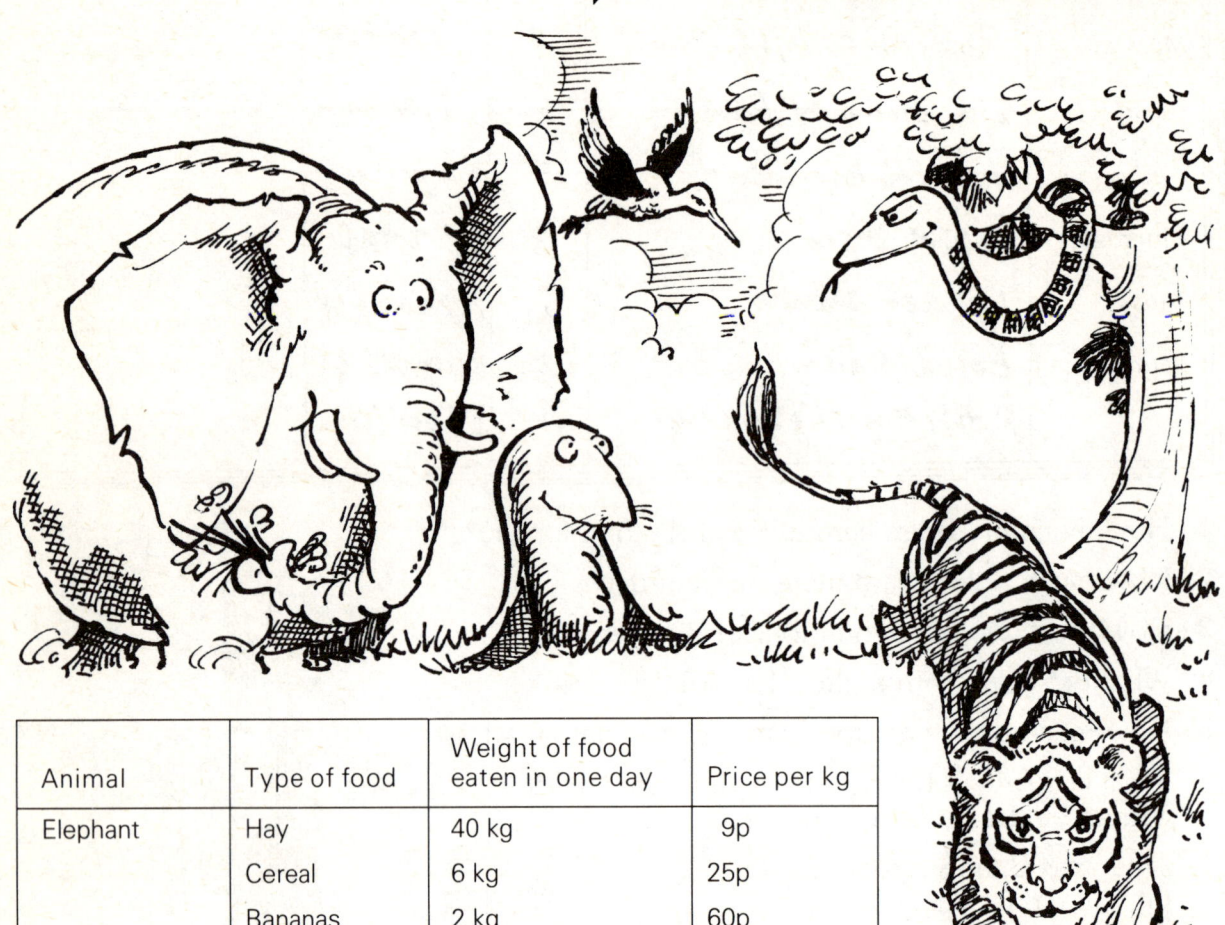

Animal	Type of food	Weight of food eaten in one day	Price per kg
Elephant	Hay	40 kg	9p
	Cereal	6 kg	25p
	Bananas	2 kg	60p
	Onions	2 kg	8p
	Carrots	1 kg	30p
	Oranges	1 kg	50p
Tiger	Meat	7 kg (starves one day)	30p
Indian python	Chickens	2 a week	30p each
Grey seal	Mackerel	6 kg	50p
White stork	Mince	1 kg	30p
	Sprats	1 kg	40p

1. How much hay does one elephant eat in a week?
2. How much cereal does one elephant eat in a week?
3. What weight of bananas does one elephant eat in a week?
4. What weight of onions does one elephant eat in a week?
5. What weight of carrots does one elephant eat in a week?
6. What weight of oranges does one elephant eat in a week?
7. How much meat does one tiger eat in a week? (**Think!**)
8. How many chickens does one python eat in a week? (**Think!**)
9. How much mackerel does one grey seal eat in a week?
10. How much mince does one white stork eat in a week?
11. What weight of sprats does one white stork eat in a week?

Harder questions

12. Find the cost of feeding one elephant for a week.
13. Find the cost of feeding one tiger for a week.
14. Find the cost of feeding a python for a week.
15. Find the cost of feeding one grey seal for a week.
16. Find the cost of feeding one white stork for a week.

Even harder questions

At the time this book was written, Flamingo Land Zoo had the following animals:
2 elephants   5 tigers   5 Indian pythons
2 grey seals   8 white storks

17. Find the weight of food needed for a week to feed
    (a) all the elephants
    (b) all the seals
    (c) all the storks
    (d) all the pythons
    (e) all the tigers
18. What is the total cost of feeding each of these groups of animals for one week?

16 Battery hens

The number on each cage is the number of eggs that each hen will lay in one week. Make a list of the number of eggs that each hen will lay in one day.
Example: 21 ÷ 7 = 3   Jemma will lay 3 eggs a day.

1. An egg box holds 6 eggs. How many egg boxes will be needed for Lucy's eggs per week?
2. If the farmer sells his eggs for 30p a box, how much will he get for Lucy's eggs in one week?
3. How much will he get for Henrietta's eggs per week? (Count full boxes only.)
4. How much will he get for Juno's eggs per week (full boxes only)?
5. How much will he get for Freda's eggs per week (full boxes only)?
6. How much will he get for Jemma's eggs per week (full boxes only)?

17 Wordwalls

Write out the answers to these questions in **words**.
Write them on squared paper, set out in a wall as shown.
The first letters of your answers will then form the answers to the riddles.

1. What happened when the cavalry lost their horses?

   They were –

2. How did the scouts sleep when some of their equipment blew away?

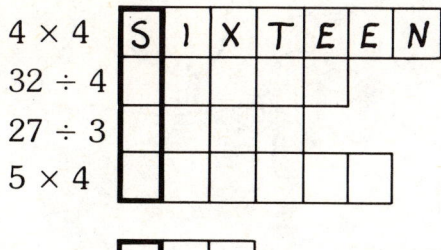

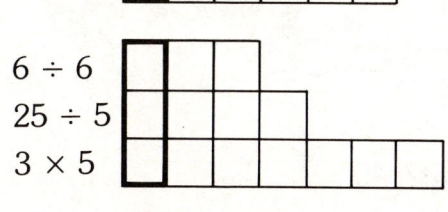

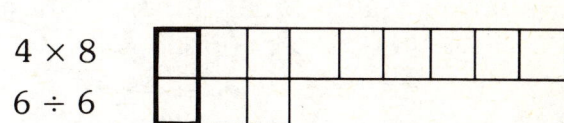

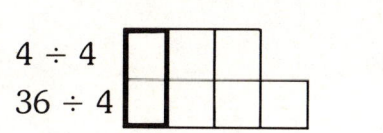

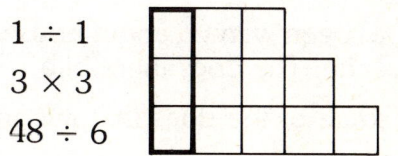

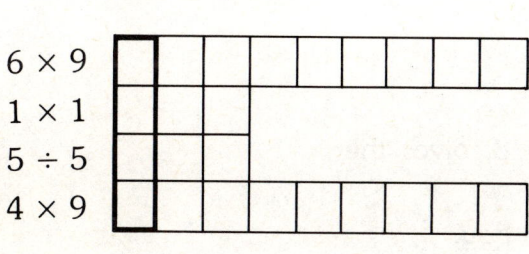

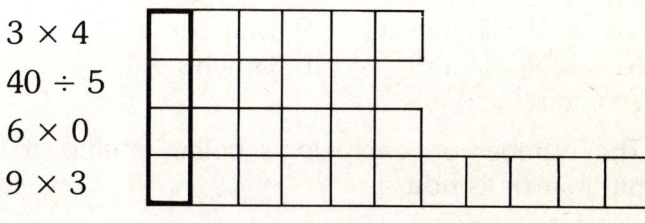

18 Run, postman, run

The new postman has been warned about houses number 32, 40 and 64 in Larch Avenue. The dogs there bite.

The picture shows some of the dogs that live in Larch Avenue. They are:

4 Dalmatian	8 Dachshund
5 Irish Wolf Hound	9 Bull Terrier
6 Bassett Hound	10 Greyhound
7 Yorkie	

The number on each dog's collar, multiplied by 8, gives the number of its house.

1. Make a list to show the house number of each dog.
   Example: $4 \times 8 = 32$ The Dalmatian lives at number 32.

2. Which dogs should the postman avoid?

3. The houses in Larch Avenue are numbered from 1 to 86. There are three more dogs in Larch Avenue. They live at houses whose numbers divide exactly by 8. Give the numbers of the houses where these dogs live.

4. What numbers will the dogs have on their collars?

19 The Mad Inventor

The Mad Inventor has imprisoned Lucy, our heroine, in a booby-trapped room. If she steps only on squares that divide exactly by 8 she will be safe. Any other squares will plunge her into the snake pit below.

Copy the grid onto squared paper and work out a safe route for Lucy.

20 Sponsored event

Greenfields Sponsor Sheet		
Name of Pupil WILL HIGGINS		
Class 4S Laps Completed 9		
Sponsor	Amount per lap	Amount collected
Mr. Higgins	10p	
Mrs Higgins	5p	
Jane Higgins	1p	
S. Chappel	3p	
H. Chappel	8p	
G. Binns	6p	
M. A. Binns	4p	
A. Sommerfield	9p	
J. Sommerfield	2p	
H. Pickersgill	7p	
M Pickersgill	8p	
B. Moore	6p	
F. Richardson	9p	
K. E. White	7p	
	Total	

1. Copy the list of Will's sponsors and work out how much he must collect from each person.
2. How much did he collect altogether?
3. Make out a sponsor sheet for yourself and ask your friends to sponsor you (not more than 10p per lap).
4. How much would you be able to collect from each person if you managed the same number of laps as Will?

21 The School Play

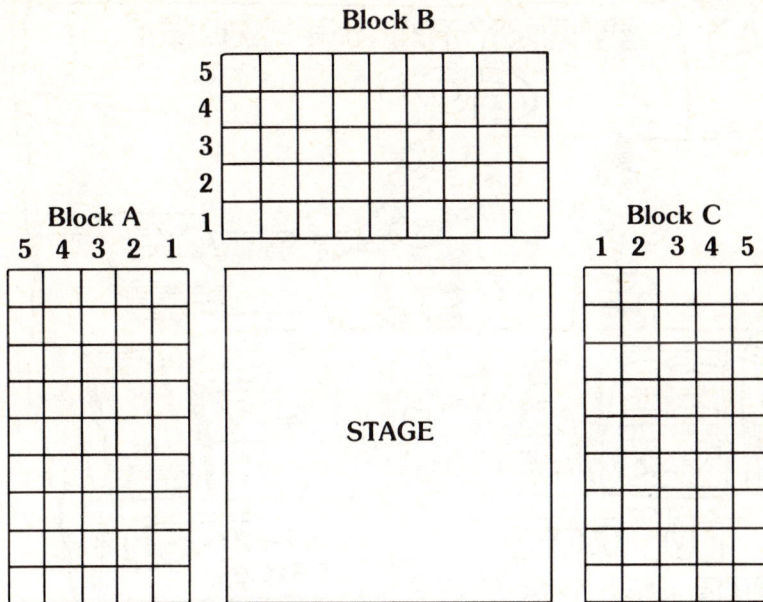

This is the seating plan for the school play. There are five rows in each block and nine seats in a row.

1. On Monday forty-five seats were occupied in block B. How many rows were filled?

2. On Tuesday sixty-three children sat in the front rows. How many rows in each block were filled?

3. On Wednesday three large parties booked to see the play. Decide how many rows each group needed and which block to put them in.
   (a) 45 old age pensioners
   (b) The Mayor and his party – 18 people altogether
   (c) 54 staff and pupils from a neighbouring school

4. On Thursday twenty-seven people booked seats in block B. The rest of the seats were occupied by friends of the Head. How many friends came?

5. Eighty-one seats were empty on Friday. How many rows was this?

6. Tickets were £1 each. £72 was taken on Monday. How many full rows were occupied?

7. How much money was taken on Friday?

22 10p Sale

Find out how much each person spent. Show all your working out.

1. Jim bought 2 tennis balls, 3 notebooks and a ruler.
2. Helen bought a Kit-Kat, 6 giant gobstoppers, 2 packets of crisps and 3 pencils.
3. Ryan bought 7 notebooks, 6 pencils, 9 pens and a ruler.
4. Alan bought 8 packets of crisps, 6 tennis balls and 5 notebooks.
5. Adriana bought 2 rulers, 5 Kit-Kats and a pen.
6. At the end of the sale there were 8 tennis balls, 7 rulers, 10 pens, 3 packets of crisps and 9 giant gobstoppers left. Mr. Wiseman bought the lot. Make a list showing how much he spent on each item.
7. How much did Mr. Wiseman spend altogether?

23 The Town Band

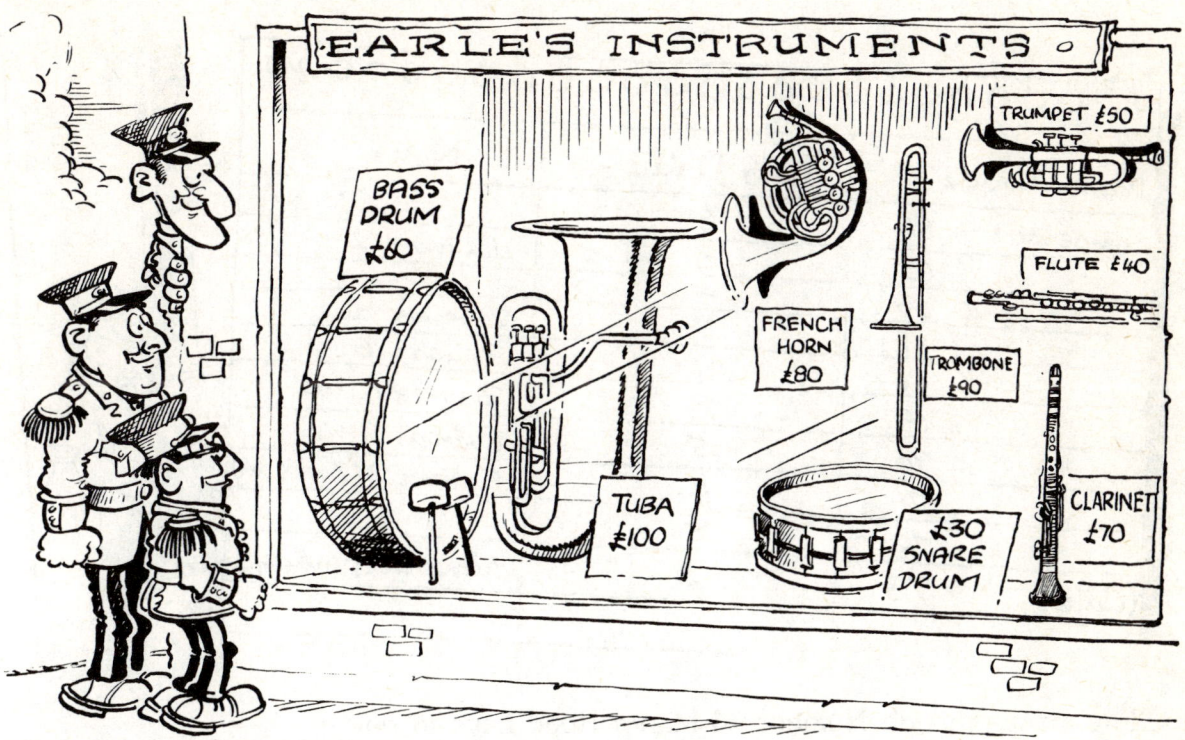

The Town Band have had their instruments stolen. It is only ten weeks until the championship. Mrs. Earle has said they may use her second-hand instruments to practise on while they are paying for them in ten weekly instalments.

The band need a tuba, a trumpet, a flute, a clarinet, a snare drum, a trombone, a French horn and a bass drum.

1. Make a list showing the weekly payments on each instrument.
   Example: The trumpet costs £50.   50 ÷ 10 = 5
   This is £5 a week.

Ten members of the Junior band want to go to watch the championship.
The total cost will be:

Cost of mini bus	£100
Overnight accommodation	£ 90
Entrance fee for competition	£ 50
Lunch	£ 30

2. Make out a list of the cost to each boy for these four items.

3. What is the total cost to each boy?

24 Three pints today, please!

Manor Road		Pints Daily	Notes
House no.	64	6	
"	65	3	Extras: 2
"	66	—	—
"	67	7	2 weeks due
"	68	9	Extras: 6
"	69	2	3 weeks due
"	70	5	Extras: 2 + 2
"	71	4	Extras: 2 (2wks due)
"	72	8	
(Café)	73	10	None on Sundays

1. How many pints do Number 64 Manor Road have in one week?
2. Number 65 had 2 extra pints this week. How many pints did they have altogether?
3. Number 67 owe for last week's milk as well as this week's. How many pints do they have to pay for this week?
4. Number 68 is a hostel. They had extra milk for a party on Friday. How many pints did they have this week?
5. Number 69 owe for three weeks' milk. How many pints are to be paid for?
6. How many pints have number 70 had? (Don't forget to look for extras.)
7. How many pints do number 71 owe for?
8. How many pints do number 72 have in one week?
9. Next week number 72 will be away for two days. How many pints will they have?
10. Number 73 is a cafe which closes on a Sunday. How many pints of milk do they normally have in one week?

25 Assault course

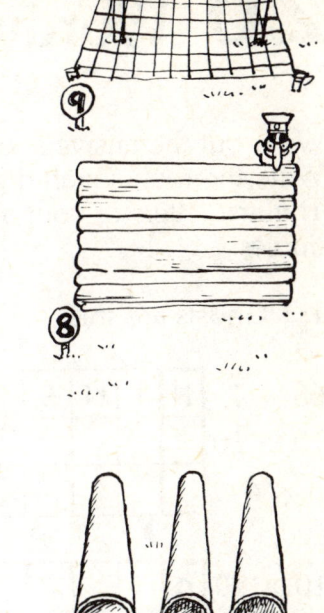

Our three raw recruits are waiting to set off on their first assault course. Multiply the number on their back by the number of each obstacle to discover the points that they score.

Copy and complete the table below to see who is the best recruit.

Note: If a score ends in 2, then the boy has failed that obstacle. Write 'fail' in place of a score

	No. 7 Ralph		No. 8 Roy		No. 9 Roger	
Obstacle 3	3×7=21	21				
Obstacle 4			4×8=32	fail		
Obstacle 5						
Obstacle 6						
Obstacle 7						
Obstacle 8						
Obstacle 9						
Total points						

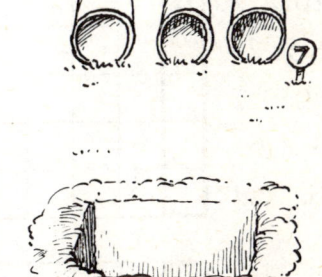

1. Who was the top recruit?
2. Who was second best?
3. Who had most failures?
4. The worst recruit was sent home. Who was he?

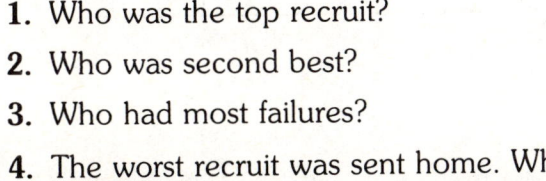

26 Wordwalls

Write out the answers to these questions in **words**.
Write them on squared paper, set out in a wall as shown.
The first letters of your answers will then form the answers to the riddles.

1. Ghosts are this:

63 ÷ 7
10 ÷ 10
5 × 4

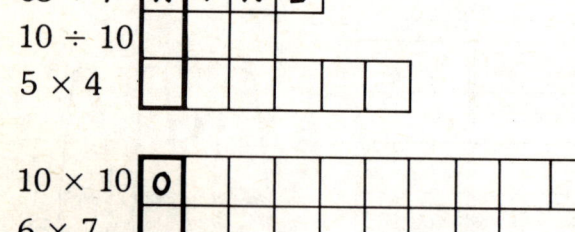

10 × 10
6 × 7
70 ÷ 7
3 × 6
10 × 9

9 × 8
10 × 8
24 ÷ 3
9 × 0

2. Don't eat this if you wear false teeth!

8 × 9
9 ÷ 9
7 × 2
6 × 6

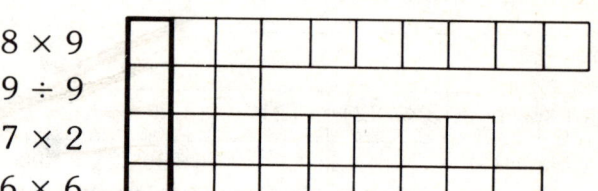

9 × 3
10 × 10
7 × 6
7 × 8
72 ÷ 9
9 × 9

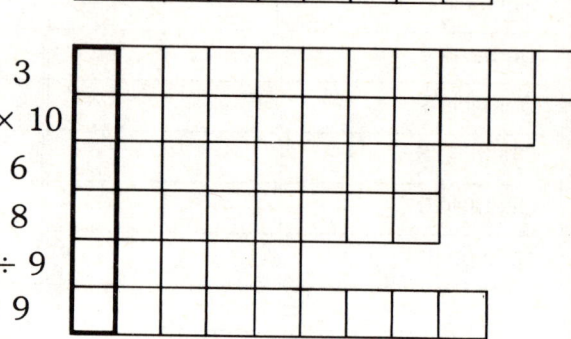

27 Coded messages

Here is a message in code. Work out the sum in each bracket, then use the code to find what letter it represents. Show all your working out.

Code

A	B	C	D	E	F	G	H	I	J	K	L	M
1	2	3	4	5	6	7	8	9	10	11	12	13

N	O	P	Q	R	S	T	U	V	W	X	Y	Z
14	15	16	17	18	19	20	21	22	23	24	25	26

Message 1

(5×4) $(35 \div 7)$ $(3 \times 6 + 1)$ (5×4)
(3×4) $(9 \div 9)$ (7×3) (7×2) $(27 \div 9)$ $(40 \div 5)$
$(8 \div 8)$ (2×10)
(7×2) (5×3) (3×5) (2×7)
(5×4) $(10 \div 10)$ $(3 \times 4 - 1)$ $(40 \div 8)$
$(27 \div 3)$ $(2 \times 7 - 1)$ $(3 \times 5 - 2)$ $(20 \div 4)$ $(24 \div 6)$ $(45 \div 5)$ $(9 \div 9)$ (2×10) $(45 \div 9)$
$(6 \div 6)$ $(24 \div 8)$ (4×5) $(54 \div 6)$ (5×3) (7×2)

Here is a message in code. Work out the sum in each bracket, then use the code to find what letter it represents. Show all your working out.

Code

A	B	C	D	E	F	G	H	I	J	K	L	M
1	2	3	4	5	6	7	8	9	10	11	12	13

N	O	P	Q	R	S	T	U	V	W	X	Y	Z
14	15	16	17	18	19	20	21	22	23	24	25	26

Message 2

$(3 \times 4 + 1)$ $(20 \div 4)$ $(40 \div 8)$ (5×4)
(2×10) $(24 \div 3)$ $(30 \div 6)$
(10×2) (3×6) $(7 \div 7)$ (3×3) (2×7)
$(27 \div 3)$ (7×2)
(3×4) (7×3) $(2 \times 5 + 3)$ $(3 \times 6 + 1)$ (2×2) (5×1) (7×2)
$(36 \div 6)$ (6×3) $(36 \div 4)$ $(12 \div 3)$ $(5 \div 5)$ (5×5)
$(25 \div 5)$ $(2 \times 10 + 2)$ $(40 \div 8)$ (2×7) $(45 \div 5)$ (7×2) $(21 \div 3)$

$(4 \div 4)$ $(40 \div 10)$ $(3 \times 7 + 1)$ $(36 \div 4)$ $(6 \times 3 + 1)$ $(25 \div 5)$
(7×3) (5×4) $(3 \times 4 + 1)$ (3×5) $(9 \times 2 + 1)$ (2×10)
$(27 \div 9)$ $(9 \div 9)$ (3×7) (4×5) (9×1) (5×3) (7×2)

$(7 \times 3 + 2)$ $(15 \div 3)$
$(64 \div 8)$ $(8 \div 8)$ $(7 \times 3 + 1)$ $(35 \div 7)$
$(14 \div 7)$ $(10 \div 2)$ $(45 \div 9)$ (2×7)
(3×5) $(16 \div 8)$ $(3 \times 6 + 1)$ $(50 \div 10)$ (9×2) $(3 \times 7 + 1)$ $(20 \div 4)$ $(20 \div 5)$

Now write a message for a friend to decode.

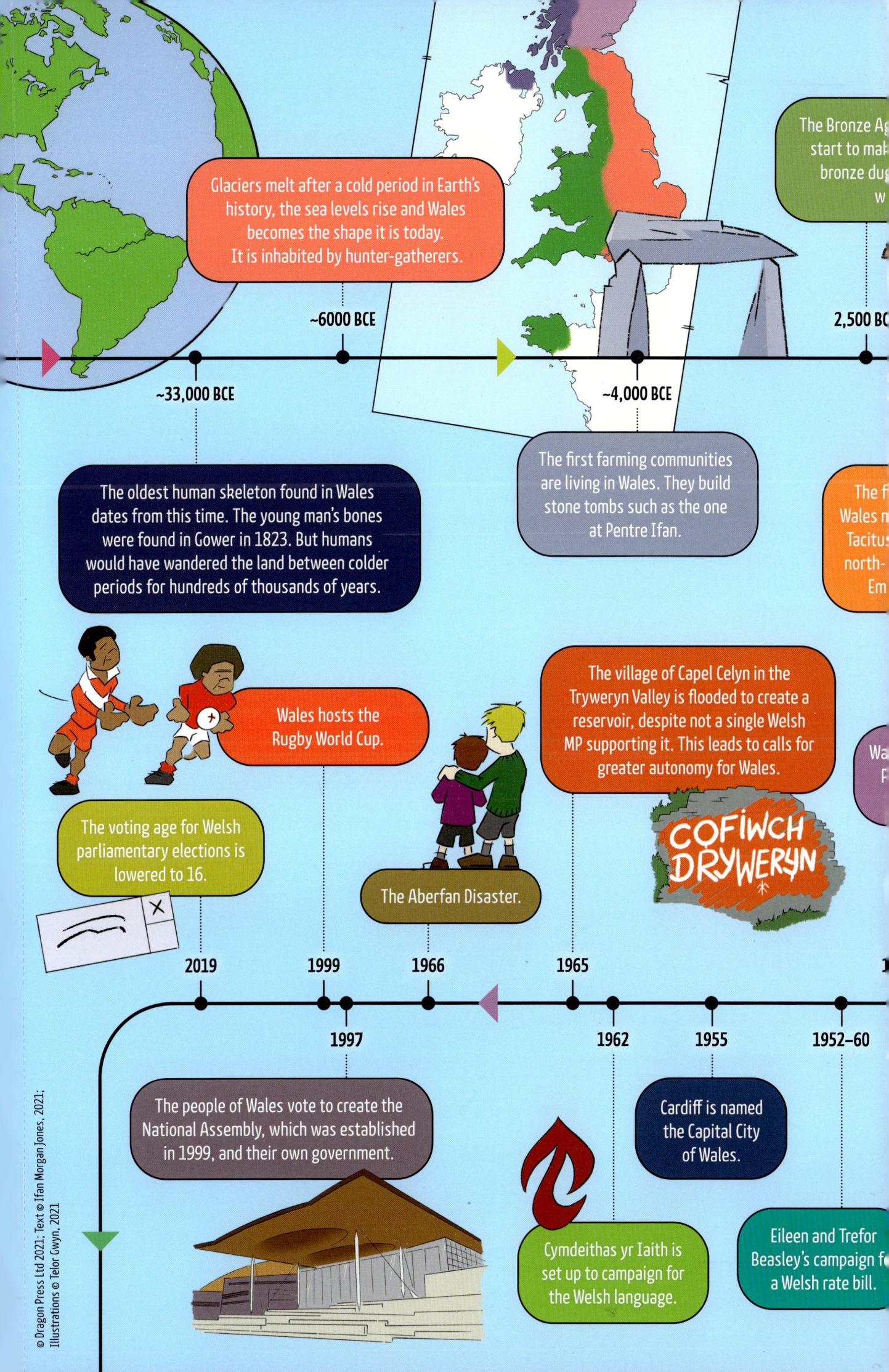

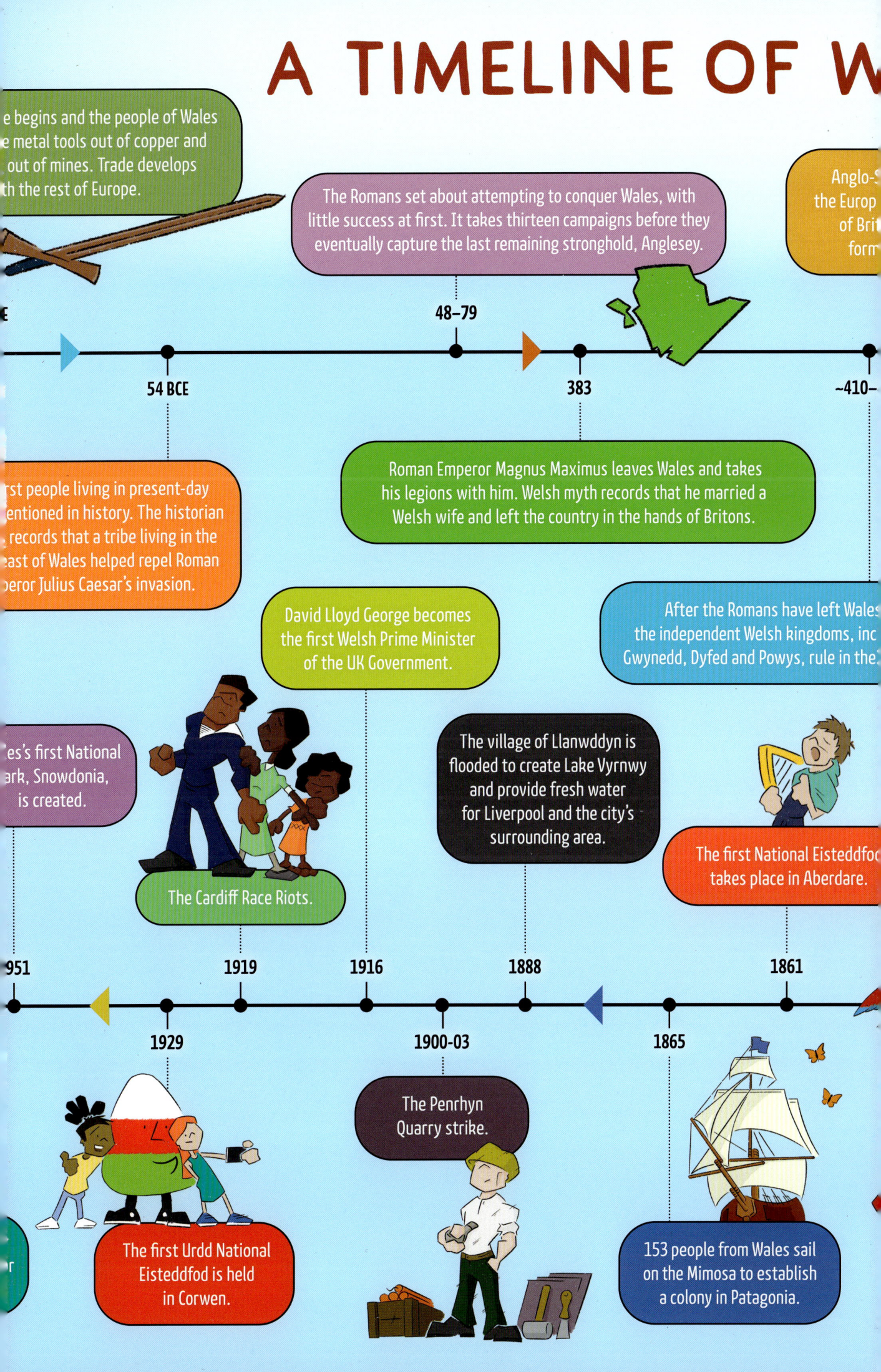

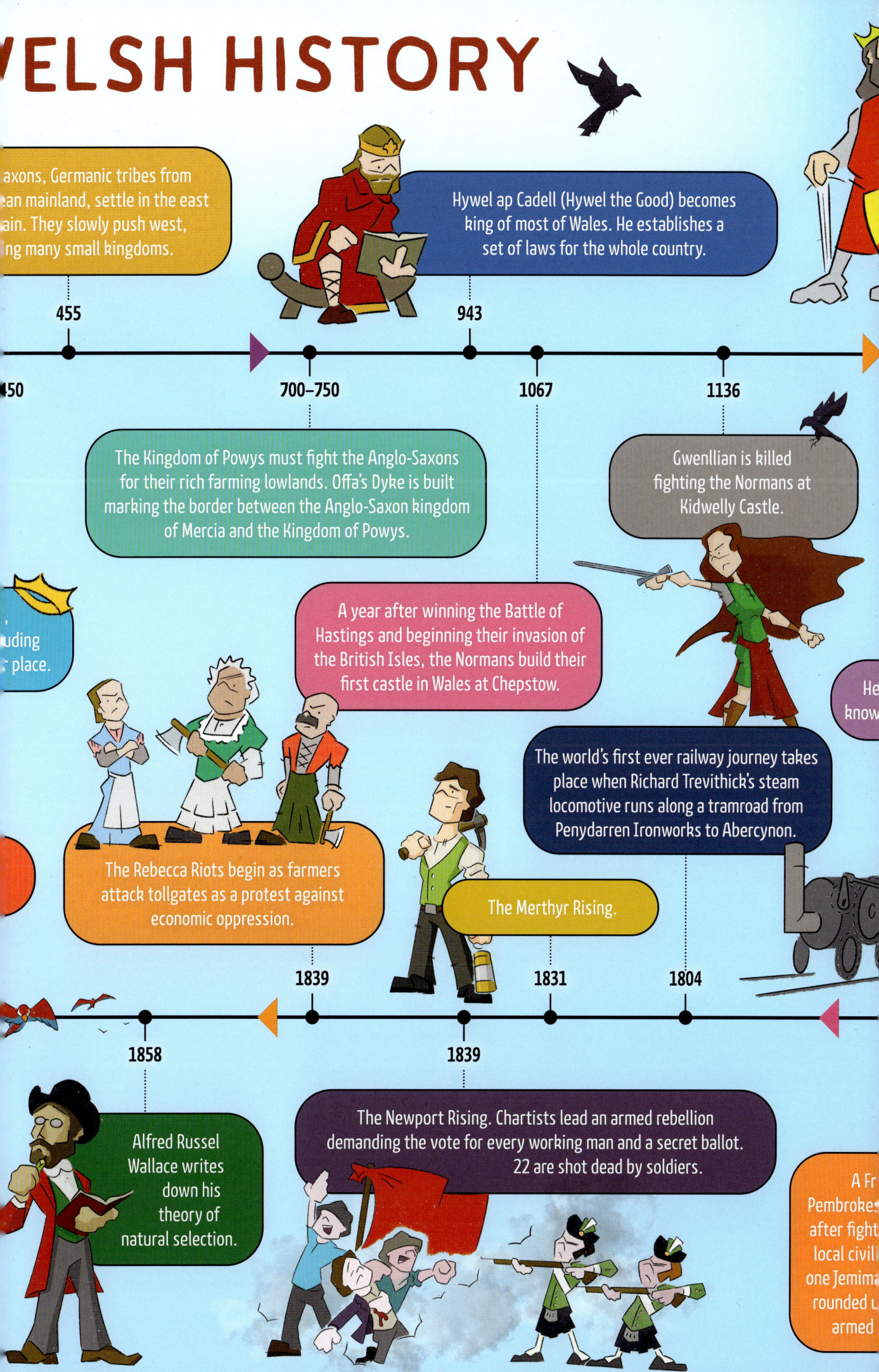

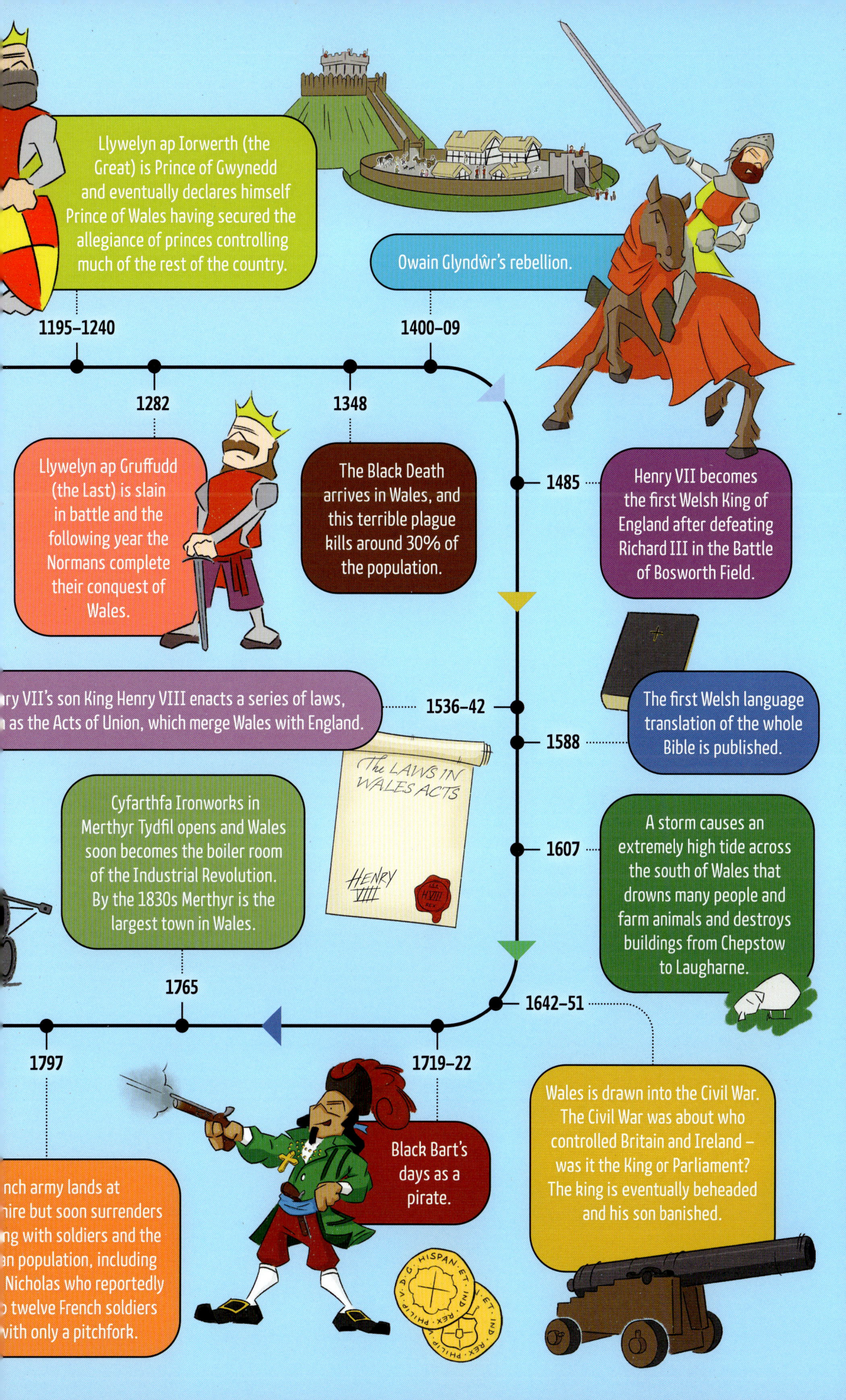